Contents

Executive Summary

Mobile technology has come a long way in the last quarter century. In the 1980s, mobile cellular devices weighed over two pounds, were about the size of a brick, and cost close to $1,000. And they could be used for one thing: phone calls.

Fast forward to 2012: In the fourth quarter of 2012, consumers worldwide bought approximately 217 million smartphones, which weigh about four ounces and fit in the palm of the hand. Consumers derive enormous benefits from these devices, which are used to make audio and video phone calls, buy movie tickets, check traffic on the regular commute, browse a digital library while waiting for an appointment, and connect with friends for spontaneous get-togethers.

The mobile ecosystem has changed in other ways, too. In the 1980s, the companies profiting from mobile devices were those that manufactured the devices and provided cellular service. Today, the cast is much bigger, and includes operating systems (*e.g.,* Google, Apple, Amazon, Microsoft, Blackberry), developers of applications, and advertising networks.

The complexity of the ecosystem raises 21st century concerns: When people use their mobile devices, they are sharing information about their daily lives with a multitude of players. How many companies are privy to this information? How often do they access such content and how do they use it or share it? What do consumers understand about who is getting their information and how they are using it?

The Federal Trade Commission ("FTC" or "Commission") has worked on privacy issues for more than forty years, and in 2000 began considering the privacy implications raised by consumers' growing use of mobile devices. Most recently, in May 2012, the FTC hosted a mobile privacy panel discussion that focused on transparency: With so many players collecting and using consumer data, who should provide privacy information to consumers? Given the limited screen space of mobile devices, how can this information be conveyed?

Based on the Commission's prior work in this area, the panel discussions, and the written submissions, this report offers several suggestions for the major participants in the mobile ecosystem as they work to improve mobile privacy disclosures.

- **Platforms, or operating system providers** offer app developers and others access to substantial amounts of user data from mobile devices (*e.g.,* geolocation information, contact lists, calendar information, photos, etc.) through their application programming interfaces (APIs). In addition, the app stores they offer are the interface between users and hundreds of thousands of apps. As a result, platforms have an important role to play in conveying privacy information to consumers. While some platforms have already implemented some of the recommendations below, those that have not should:

- Provide just-in-time disclosures to consumers and obtain their affirmative express consent before allowing apps to access sensitive content like geolocation;

- Consider providing just-in-time disclosures and obtaining affirmative express consent for other content that consumers would find sensitive in many contexts, such as contacts, photos, calendar entries, or the recording of audio or video content;

- Consider developing a one-stop "dashboard" approach to allow consumers to review the types of content accessed by the apps they have downloaded;

- Consider developing icons to depict the transmission of user data;

- Promote app developer best practices. For example, platforms can require developers to make privacy disclosures, reasonably enforce these requirements, and educate app developers;

- Consider providing consumers with clear disclosures about the extent to which platforms review apps prior to making them available for download in the app stores and conduct compliance checks after the apps have been placed in the app stores;

- Consider offering a Do Not Track (DNT) mechanism for smartphone users. A mobile DNT mechanism, which a majority of the Commission has endorsed, would allow consumers to choose to prevent tracking by ad networks or other third parties as they navigate among apps on their phones.

- **App developers** should:

  - Have a privacy policy and make sure it is easily accessible through the app stores;

  - Provide just-in-time disclosures and obtain affirmative express consent before collecting and sharing sensitive information (to the extent the platforms have not already provided such disclosures and obtained such consent);

  - Improve coordination and communication with ad networks and other third parties, such as analytics companies, that provide services for apps so the app developers can provide accurate disclosures to consumers. For example, app developers often integrate third-party code to facilitate advertising or analytics within an app with little understanding of what information the third party is collecting and how it is being used. App developers need to better understand the software they are using through improved coordination and communication with ad networks and other third parties.

- Consider participating in self-regulatory programs, trade associations, and industry organizations, which can provide guidance on how to make uniform, short-form privacy disclosures.

Advertising networks and other third parties should:

- Communicate with app developers so that the developers can provide truthful disclosures to consumers;

- Work with platforms to ensure effective implementation of DNT for mobile.

App developer trade associations, along with academics, usability experts and privacy researchers can:

- Develop short form disclosures for app developers;

- Promote standardized app developer privacy policies that will enable consumers to compare data practices across apps;

- Educate app developers on privacy issues.

Many companies in the mobile ecosystem have already begun addressing the challenge of developing effective privacy disclosures, and FTC staff applauds these efforts. The National Telecommunications and Information Agency ("NTIA"), within the U.S. Department of Commerce, has initiated a multistakeholder process to develop a code of conduct on mobile application transparency. To the extent that strong privacy codes are developed, the FTC will view adherence to such codes favorably in connection with its law enforcement work. Staff hopes that this report will provide important input for all participants in that process as well as stakeholders developing guidance and initiatives in this area.

Mobile Privacy Disclosures: Building Trust Through Transparency

The FTC has been examining the privacy implications of mobile devices since 2000, in the form of workshops, law enforcement actions, consumer education, testimony, and policy reports. On May 30, 2012, the FTC held a workshop entitled "In Short: Advertising and Privacy Disclosures in a Digital World." The concluding panel at that workshop explored how privacy disclosures on mobile devices could be short, effective, and accessible to consumers.[1]

Based on more than a decade of work on mobile privacy issues and recent data obtained through panel discussions and comments, the Commission offers this staff report providing recommendations for best practices on mobile privacy disclosures.[2] First, the report reviews the benefits and privacy risks of mobile technologies. Second, it discusses the FTC's efforts to address mobile privacy, as well as its research on disclosure issues generally. It then summarizes general themes raised by panel participants. Finally, it sets forth recommendations for best practices to key commercial players involved in the mobile arena – platforms,[3] app developers, third parties such as ad networks and analytics companies, and trade associations.[4] The recommendations are intended to promote more effective privacy disclosures.

---

1. The workshop also addressed advertising disclosure challenges that have emerged since the FTC first issued its online advertising disclosure guidance, "Dot Com Disclosures," approximately 12 years ago. The FTC expects to issue in a separate report updated guidance addressing advertising disclosures.

2. Commission staff recognizes that disclosure is only one element of privacy protection on mobile devices and that mobile companies should consider privacy issues at every stage of product development (i.e., adopt "privacy by design") by implementing substantive protections, such as data minimization and data security, as well as procedural safeguards aimed at integrating those substantive protections into a company's everyday business operations. *See* FTC, *Protecting Consumer Privacy in an Era of Rapid Change, Recommendations for Businesses and Policymakers* (Mar. 2012), *available at* http://www.ftc.gov/os/2012/03/120326privacyreport.pdf. These issues are beyond the scope of this report and not addressed substantively here; however, this in no way minimizes their importance.

3. In this report, the term "platform" refers to mobile operating systems, such as Apple's iOS, Google's Android, RIM's BlackBerry OS, and Microsoft's Windows Phone, along with the app stores they offer, such as the Apple App Store, Google Play, BlackBerry App World, and Microsoft's Windows Store.

4. Other mobile ecosystem participants – such as carriers, handset manufacturers, and chip makers – also should review these recommendations carefully and consider how they may contribute to improving mobile privacy disclosures.

I. Benefits and Risks of Mobile Technologies

A decade ago, smartphones and tablets did not exist; today, they are everywhere.[5] Indeed, as of 2012, smartphone owners represented a majority of U.S. mobile subscribers. In addition, ten years ago, the term "app" had not entered common parlance; today, there are over 800,000 available in the Apple App Store and 700,000 on Google Play.[6] In less than three years, a "tablet" has become as synonymous with an iPad or Kindle Fire as with a medication. In the coming years, consumers' use of smartphones, tablets, and other mobile devices is projected to grow at staggering rates.[7]

No one doubts that the rapid growth of mobile technologies provides enormous value to both businesses and consumers. Mobile devices are revolutionizing how consumers interact, communicate, and carry out everyday activities. In a typical day a consumer may use a mobile device to read the latest news, email, text, pay bills, place and receive phone calls, post status updates on a social networking site, download and launch an app to find nearby movie theaters and buy tickets to the latest release, and even pay for a cup of coffee.

At the same time, mobile technology presents unique privacy challenges. First, more than other types of technology, mobile devices are typically personal to an individual, almost always on, and with the user. This can facilitate unprecedented amounts of data collection. The data collected can reveal sensitive information, such as communications with contacts, search queries about health conditions, political interests, and other affiliations, as well as other highly personal information.[8] This data also may be shared with third parties, for example, to send consumers behaviorally targeted advertisements.

Second, in the complicated mobile ecosystem, a single mobile device can facilitate data collection and sharing among many entities, including wireless providers, mobile operating

---

5. *See, e.g.,* Nielsen Wire, *America's New Mobile Majority: a Look at Smartphone Owners in the U.S.* (May 7, 2012), available at http://blog.nielsen.com/nielsenwire/online_mobile/who-owns-smartphones-in-the-us.

6. *See* Press Release, Apple, *Apple Updates iOS to 6.1* (Jan. 28, 2013), *available at* https://www.apple.com/pr/library/2013/01/28Apple-Updates-iOS-to-6-1.html; Business Week, *Google Says 700,000 Applications Available for Android* (Oct. 29, 2012), *available at* http://www.businessweek.com/news/2012-10-29/google-says-700-000-applications-available-for-android-devices.

7. *See, e.g.,* AppNation & Rubinson Partners, *How Big is the US App Economy? Estimates and Forecasts 2011-2015* (2011), *available* at http://www.slideshare.net/joelrubinson/an3-us-appeconomy20112015 (projecting that number of U.S. consumers owning smartphones will double between 2011 and 2015, and that the number of U.S. consumers owning tablets will more than triple).

8. Jennifer M. Urban et al., *Mobile Phones and Privacy* (July 11, 2012) at 5, *available at* http://ssrn.com/abstract=2103405.

system providers, handset manufacturers, application developers, analytics companies, and advertisers to a degree unprecedented in the desktop environment. This can leave consumers wondering where they should turn if they have questions about their privacy.

Third, mobile devices can reveal precise information about a user's location that could be used to build detailed profiles of consumer movements over time and in ways not anticipated by consumers. Indeed, companies can use a mobile device to collect data over time and "reveal[] the habits and patterns that mark the distinction between a day in the life and a way of life."[9] Even if a company does not intend to use data in this way, if the data falls in the wrong hands, the data can be misused and subject consumers to harms such as stalking or identity theft.[10]

In recent studies, consumers have expressed concern about their privacy on mobile devices. For example, a nationwide survey indicated that 57% of all app users have either uninstalled an app over concerns about having to share their personal information, or declined to install an app in the first place for similar reasons.[11] Similarly, in a 2011 survey of U.S. smartphone users, less than one-third of survey respondents reported feeling in control of their personal information on their mobile devices.[12] Lack of attention to these concerns could lead to an erosion of trust in the mobile marketplace, which could be detrimental to both consumers and industry.[13]

Finally, with many devices possessing screens of just a few inches, there are practical challenges in terms of how critical information – such as data collection, sharing of information, and use of geolocation data – is conveyed to consumers.

---

9. *United States v. Maynard*, 615 F.3d 544, 562 (D.C. Cir. 2010).

10. *See* Government Accountability Office, *Mobile Device Location Data: Additional Federal Actions Could Held Protect Consumer Privacy* (Sept. 2012), *available at* http://www.gao.gov/assets/650/648044.pdf.

11. Pew Internet & American Life Project, *Privacy and Data Management on Mobile Devices* (Sept. 5, 2012), *available at* http://pewinternet.org/ ~ /media//Files/Reports/2012/PIP_MobilePrivacyManagement.pdf. Consistent with these findings, a recent study by the Berkeley Center for Law and Technology revealed that most consumers consider the information on their mobile devices to be private. *See* Jennifer M. Urban et al., *Mobile Phones and Privacy*, *supra* note 8, at 2.

12. TRUSTe, *Mobile Privacy Survey Results* (last visited Jan. 28, 2013), *available at* http://www.truste.com/ why_TRUSTe_privacy_services/harris-mobile-survey/.

13. *See Protecting Mobile Privacy: Your Smartphones, Tablets, Cell Phones and Your Privacy: Hearing Before the Subcomm. for Privacy, Technology and the Law of the S. Comm. on the Judiciary*, 112th Cong. (2011) (statement of Alan Davidson, Director of Public Policy, Google Inc.) ("If we fail to offer clear, usable privacy controls, transparency in our privacy practices, and strong security, our users will simply switch to another provider. This is as true for our services that are available on mobile devices as it is for those that are available on desktop computers.").

II. FTC's Prior Work

A. Efforts to Address Mobile Privacy

The FTC has been engaged in a multi-pronged effort to understand and address mobile privacy concerns. Since 2000, the Commission has hosted numerous workshops and outreach events to develop an understanding of mobile technology.[14] More recently, the FTC assembled a Mobile Technology Unit that conducts research, monitors the various platforms, app stores, and applications, and trains FTC staff on mobile issues. Through this Unit, the Commission is ensuring that it has the necessary technical expertise, understanding of the marketplace, and tools to monitor, investigate, and prosecute deceptive and unfair practices in the mobile arena.

The Commission has identified three key areas for ongoing work to address mobile privacy concerns: enforcement, outreach, and policy initiatives.

First, the Commission continues to bring enforcement actions against companies operating in the mobile environment. Most recently, the Commission settled charges that a social networking service deceived consumers regarding the collection of their address book information through its mobile application, and illegally collected information from children under age 13 without providing notice and obtaining parental consent, in violation of the Children's Online Privacy Protection Act (COPPA).[15] In another action, the Commission brought charges against a peer-to-peer file-sharing application developer whose software likely would cause consumers to unwittingly expose sensitive personal files stored on their mobile devices.[16] Futher, in the Commission's first Fair Credit Reporting Act (FCRA) case involving a mobile app, an enterprise that compiled and sold criminal record reports settled charges

---

14. *See, e.g.*, FTC Workshop, *The Mobile Wireless Web, Data Services and Beyond: Emerging Technologies and Consumer Issues* (Dec. 11-12, 2000), *available at* www.ftc.gov/bcp/workshops/ wireless/index.shtml; FTC Workshop, *Protecting Consumers in the Next Tech-ade* (Nov. 6-8, 2006), *available at* www.ftc.gov/bcp/workshops/techade; FTC Workshop, *Beyond Voice: Mapping the Mobile Marketplace* (May 6-7, 2008), *available at* http://www.ftc.gov/bcp/workshops/mobilemarket/index.shtml; FTC Workshop, *Pay on the Go: Consumers and Contactless Payment* (July 24, 2008), *available at* http://www.ftc.gov/bcp/workshops/payonthego/index.shtml; FTC Workshop, *Transatlantic RFID Workshop on Consumer Privacy and Data Security* (Sept. 23, 2008), *available at* www.ftc.gov/bcp/workshops/ transatlantic/index.shtml; FTC Workshop, *Paper, Plastic... or Mobile? An FTC Workshop on Mobile Payments* (Apr. 26, 2012), *available at* www.ftc.gov/bcp/workshops/mobilepayments.

15. *See United States v. Path, Inc.*, No. C13-0448 (N.D. Cal. Jan. 31, 2013) (proposed consent order), *available at* http://www.ftc.gov/os/caselist/1223158/index.shtm.

16. *See In the Matter of Frostwire LLC*, No. 1:11-cv-23643 (S.D. Fla. Oct. 12, 2011) (consent order), *available at* http://www.ftc.gov/os/caselist/1123041/index.shtm.

that it operated as a consumer reporting agency without taking consumer protection measures required by law.[17]

Second, the Commission continues to educate consumers, businesses, and others on mobile privacy. The Commission provides consumer education materials on such topics as use of Wi-Fi networks and disposal of old cell phones in a manner that safely protects personal and sensitive information.[18] In addition, given the fact that many players in the mobile ecosystem – particularly app developers – are small businesses, the Commission has focused resources on business education as well. Most recently, the Commission issued guidance to app developers on mobile security.[19] In August 2012, the FTC published educational materials offering guidance to help mobile app developers comply with truth-in-advertising standards and basic privacy principles.[20] The FTC also actively works to educate mobile companies directly by speaking at meetings of app developers and distributing business education materials.

Third, the Commission and staff have issued several policy recommendations for mobile companies, as described below.[21]

1. Privacy Report

In March 2012, the Commission issued its privacy report ("Privacy Report") setting forth best practices for businesses to protect consumers' privacy and give them greater control

---

17. *See In the Matter of Filiquarian Publishing, LLC*, FTC File No. 112 3195 (Jan. 10, 2013) (proposed consent order), *available at* http://www.ftc.gov/os/caselist/1123195/index.shtm.

18. *See* Press Release, FTC, *FTC Offers Tips on Wise Use of Wi-Fi Networks* (Feb. 11, 2011), *available at* http://www.ftc.gov/opa/2011/02/wireless.shtm; FTC, *Disposing of Your Mobile Device* (June 2012), *available at* http://www.ftc.gov/bcp/edu/pubs/consumer/alerts/alt044.shtm.

19. FTC, *Mobile App Developers: Start with Security* (Feb. 2013), *available at* http://business.ftc.gov/documents/bus83-mobile-app-developers-start-security.

20. FTC, *Marketing Your Mobile App: Get It Right from the Start* (Aug. 2012), *available at* http://business.ftc.gov/documents/bus81-marketing-your-mobile-app.

21. The Commission also has described its work on mobile privacy issues in testimony before Congress. *See Consumer Privacy and Protection in the Mobile Marketplace: Hearing Before the S. Comm. on Commerce, Science, and Transportation*, 112th Cong. (2011) (statement of David C. Vladeck, Director, Bureau of Consumer Protection, FTC), *available at* http://www.ftc.gov/os/testimony/110519mobilemarketplace.pdf; *Protecting Mobile Privacy: Your Smartphones, Tablets, Cell Phones and Your Privacy: Hearing Before the Subcomm. for Privacy, Technology and the Law of the S. Comm. on the Judiciary*, 112th Cong. (2011) (statement of Jessica Rich, Deputy Director, Bureau of Consumer Protection, FTC), *available at* http://www.ftc.gov/os/testimony/110510mobileprivacysenate.pdf.

over the collection and use of their personal data.[22] The Privacy Report calls on companies handling consumer data to adhere to three core principles:

- **Privacy by Design:** Companies should build in privacy at every stage in developing their products.

- **Simplified Consumer Choice:** For practices not consistent with the context of a transaction or a consumer's relationship with the business, companies should provide consumers with choices at a relevant time and context.

- **Greater Transparency:** Companies should disclose details about their collection and use of consumers' information.

As the Privacy Report noted, all three of these principles apply to mobile companies. With respect to privacy by design, the Report called on companies to limit collection to data they need for a requested service or transaction.[23] For example, the Privacy Report noted that a wallpaper app or an app that offers stock quotes does not need to collect location information.[24]

As to choice and transparency, the Privacy Report noted that all companies involved in data collection and sharing through mobile devices – carriers, handset manufacturers, operating system providers, app developers, and advertisers – should work together to provide privacy disclosures and ensure that they are understandable and accessible on a small screen. The Report also called on companies to develop standard formats and terminology for privacy statements applicable to their particular industries.[25] The Commission acknowledged the challenges and complexities of providing notice in the mobile environment, and indicated that these factors increase the urgency for companies providing mobile services to develop standard notices, icons, and other disclosures that businesses can use to communicate with consumers

---

22. FTC, *Protecting Consumer Privacy in an Era of Rapid Change, Recommendations for Businesses and Policymakers, supra* note 2.

23. *Id.* at 33.

24. *Id.*

25. *Id.* at 62.

in a clear and consistent way.[26] The Commission also supported the development of a Do Not Track (DNT) mechanism for both the web and mobile environments.[27]

2. Kids App Reports

In 2012, FTC staff released two reports that surveyed mobile apps for children in Apple's App Store and Google Play ("kids app reports"). In the initial report, FTC staff analyzed 400 apps directed to kids and found that, of the apps surveyed, companies offered little or no information to parents about their privacy practices.[28] The report therefore set forth recommendations for app stores, developers, and third parties providing services within the apps aimed at providing key information to parents who download apps.[29]

The report laid out a number of actions the app stores could take to improve disclosures, including providing a designated space for standardized icons or for developers to disclose their apps' data collection practices.[30] The report called on the stores, in light of an apparent lack of enforcement, to enforce the provisions of their agreements requiring developers to disclose the information their apps collect.[31] The report also urged app developers to provide information about data practices simply and succinctly, and ad networks and other third parties collecting user information through apps to disclose their privacy practices through an easily accessible method.[32]

Following the release of the initial report, FTC staff conducted a follow-up survey of 400 children's apps available through the leading two platforms.[33] This time, in addition to examining the disclosures that apps provided about their privacy practices, the new survey tested the apps' practices and compared them to the disclosures made.

---

26. *Id.* at 63-64.

27. *See Balancing Privacy and Innovation: Does the President's Proposal Tip the Scale?: Hearing Before the Subcomm. on Commerce, Manufacturing and Trade of the H. Comm. on Energy and Commerce*, 112th Cong. (2012) (statement of Jon Leibowitz, Chairman, FTC), *available at* http://www.ftc.gov/os/testimony/1 20329privacytestimony.pdf.

28. *See* FTC Staff, *Mobile Apps for Kids: Current Privacy Disclosures are Disappointing* (Feb. 2012), *available at* http://www.ftc.gov/os/2012/02/120216mobile_apps_kids.pdf.

29. *Id.* at 3.

30. *Id.*

31. *Id.*

32. *Id.*

33. *See* FTC Staff, *Mobile Apps for Kids: Disclosures Still Not Making the Grade* (Dec. 2012), *available at* http://www.ftc.gov/os/2012/12/121210mobilekidsappreport.pdf.

The survey again found that most apps still failed to provide parents with any information about the data collected through the app.[34] The survey further showed that many apps also shared information with third parties, including advertising networks, without disclosing this fact. The follow-up report called on participants in the mobile marketplace to follow the three key principles laid out in the Privacy Report.[35] Further, the follow-up report emphasized that industry players should work together to develop accurate disclosures regarding what data is collected through kids' apps, how it will be used, with whom it will be shared, and whether the apps contain interactive features such as advertising, the ability to make in-app purchases, and links to social media.[36]

B. General Research on Disclosures

Commission staff has also worked with experts to conduct consumer testing of disclosures in many contexts. Highlights of this work include the following:

Study on Financial Privacy Notices: The Gramm-Leach-Bliley Act (GLBA) requires financial companies to provide their customers with annual privacy notices. The notices that financial companies first mailed to consumers after the law was enacted were often lengthy, varied in how they informed consumers of their rights, and were generally confusing to consumers. In 2004, FTC staff worked with other agencies that enforced GLB to develop an alternative short form financial privacy notice prototype, which was finalized, with modifications, in 2009.[37] If companies use this prototype for their privacy notices, they will be granted a safe harbor for compliance with the notice requirements of the law.

The prototype notice was the subject of extensive consumer testing, which suggested the importance of several attributes in privacy disclosures: (1) simplicity; (2) good design techniques such as tables, headings, white space, bold text, bulleted lists, and a larger font size; (3) neutrality in language and presentation; (4) context; and (5) standardization.[38]

---

34. *Id.* at 4.

35. *Id.* at 21.

36. *Id.*

37. *See* Kleimann Communication Group, Inc., *Evolution of a Prototype Financial Privacy Notice: A Report on the Form Development Project* (Feb. 28, 2006) at i, *available at* http://ftc.gov/privacy/privacyinitiatives/ ftcfinalreport060228.pdf; Press Release, FTC, *Federal Regulators Issue Final Model Privacy Form* (Nov. 17, 2009), *available at* http://www.ftc.gov/opa/2009/11/glb.shtm.

38. *See* Kleimann Communication Group, Inc., *Evolution of a Prototype Financial Privacy Notice: A Report on the Form Development Project*, *supra* note 37.

Mortgage Disclosures: In 2007, FTC staff studied the effectiveness of existing mortgage disclosures, and found that they failed to convey key mortgage costs to consumers.[39] Staff developed and tested an alternative prototype notice, which was much more successful in conveying these costs. The prototype notice included the key costs in simple, easy-to-understand language, and excluded less important or confusing information. The form was layered in that it included summary information on the first page, and more detailed information on subsequent pages.

Testing of Specific Statements: FTC staff have also conducted consumer testing of specific disclosures in other contexts, such as "up to" energy efficiency claims for windows,[40] various environmental claims,[41] and language contained in opt out notices for pre-screened offers of credit.[42]

FTC staff has drawn on this experience with disclosures generally – in addition to its work on mobile privacy – in developing the recommendations in this report.

III. Common Themes From Panel Participants and Commenters

To examine how to effectively disclose privacy practices in the mobile arena, in its May 30, 2012 workshop, the FTC convened a panel of representatives from industry, trade associations, academia, and consumer privacy groups. The panel began with a demonstration of how consumers typically download apps on two major platforms – Apple and Google – and a discussion of existing privacy disclosures on each platform.[43] The panel then highlighted several private-sector initiatives to improve mobile privacy disclosures, including an icon-based program, a privacy "badge," and privacy policy generators (discussed further below). Panelists discussed the role of platforms vis-à-vis apps, consumers' understanding of mobile

---

39. FTC Staff, *Improving Consumer Mortgage Disclosures: An Empirical Assessment of Current and Prototype Disclosure Forms* (June 2007), *available at* http://www.ftc.gov/os/2007/06/P025505MortgageDisclosureReport.pdf.

40. Manoj Hastak & Dennis Murphy, *Effects of a Bristol Windows Advertisement with an "Up To" Savings Claim on Consumer Take-Away and Beliefs* (May 2012), *available at* http://www.ftc.gov/os/2012/06/120629 bristolwindowsreport.pdf.

41. FTC Staff, *The Green Guides – Statement of Basis and Purpose* (Oct. 2012), *available at* http://www.ftc.gov/os/fedreg/2012/10/greenguidesstatement.pdf.

42. Manoj Hastak, *The Effectiveness of "Opt-Out" Disclosures in Pre-Screened Credit Card Offers* (Sept. 2004), *available at* http://www.ftc.gov/reports/prescreen/040927optoutdiscprecreenrpt.pdf.

43. Apple and Google were invited to participate as panelists at the workshop but declined.

privacy issues, and consumers' perceptions of disclosures. Following the workshop, the FTC received public comments on privacy issues from interested parties.[44]

Several key themes emerged from panelists and written commenters:

Lack of Consumer Awareness and Understanding: One theme was that consumers do not know or understand current information collection and use practices occurring on mobile devices.[45] According to one participant, because consumers are unaware that many of these practices are taking place, they do not look for options providing them with control over such practices.[46] Another participant noted that when made aware of these practices, consumers typically are surprised and view the practices as underhanded.[47] Participants noted that when disclosures are made, consumers often do not understand them.[48] As a result, participants concluded that there is a need for improved mobile privacy disclosures.

Importance of Design to Address Limitations of Small Screens: Participants discussed how to improve privacy disclosures given space limitations on mobile devices, and limited attention spans of consumers. They discussed the need to develop shorthand, consistent disclosures. Options discussed included the use of icons, short form privacy notices, and layered notices. According to participants, an important benefit of these types of standardized, shorthand disclosures is that consumers could use them to easily compare practices across apps, similar to the concept of a "nutrition label" on foods.[49]

---

44. *See* FTC, # 434; FTC Announces Final Agenda and Panelists for Workshop about Advertising and Privacy Disclosures in Online and Mobile Media; FTC Project Number P114506, *available at* http://www.ftc.gov/ os/comments/inshortworkshop/index.shtm. Of the eleven written comments submitted, only six comments addressed issues raised in connection with the mobile privacy disclosures panel.

45. *See In Short Workshop, Remarks of Ilana Westerman, Create with Context,* at 218; *In Short Workshop, Remarks of Prof. Lorrie Faith Cranor, Carnegie Mellon University,* at 262-63 (noting lack of consumer awareness and understanding regarding data collection in the desktop environment and predicting that applying current practices to the mobile environment will result in even less consumer comprehension).

46. *See In Short Workshop, Remarks of Ilana Westerman, Create with Context,* at 218.

47. *See In Short Workshop, Remarks of Prof. Lorrie Faith Cranor, Carnegie Mellon University,* at 262-63.

48. *See In Short Workshop, Remarks of Prof. Lorrie Faith Cranor, Carnegie Mellon University,* at 231-32; *In Short Workshop, Remarks of Ilana Westerman, Create with Context,* at 225-226; *In Short Workshop, Remarks of Kevin Trilli, TRUSTe* at 252.

49. *See In Short Workshop, Remarks of Jennifer King, University of California, Berkeley School of Information,* at 19; *Comment of Retail Industry Leaders* Association, at 1, *available at* http://www.ftc.gov/os/comments/ inshortworkshop/00003-83138.pdf.

Participants emphasized the importance of delivering disclosures at the appropriate time.[50] By informing consumers at an appropriate moment in time, a disclosure is likely to be of greater relevance to them. Disclosures may have little meaning for a consumer if made at one point in time, yet that same disclosure may be highly relevant if made at another point in time.[51] In this context, participants discussed the importance of just-in-time disclosures, but they also emphasized the benefits of multiple points of disclosure so that consumers can learn about information collection and use practices at a number of different points during their mobile experience.[52]

Key Role of Platforms: Participants pointed out the considerable control platforms possess over how information is conveyed to consumers and the influence they possess over app developers.[53]

Accordingly, participants pointed to a number of steps platforms could take in order to improve transparency. For example, one participant described how platforms are working to make space for certain apps to display a privacy icon.[54] Another participant suggested that platforms should implement technical mechanisms geared towards greater standardization and automation that would cut down on time spent by consumers sorting through privacy policies.[55] Yet another participant indicated that platforms should be pushing developers, perhaps as part of the application upload process, to make more deliberate decisions about data practices.[56]

Regardless of the approach, there appeared to be widespread agreement among participants that platforms can do more.

---

50. *See In Short Workshop, Remarks of Pam Dixon, World Privacy Forum,* at 233-34.

51. *See In Short Workshop, Remarks of Pam Dixon, World Privacy Forum,* at 233-35; *In Short Workshop, Remarks of Ilana Westerman, Create with Context,* at 235.

52. *See In Short Workshop, Remarks of Jennifer King, University of California, Berkeley School of Information,* at 15-16; *In Short Workshop, Remarks of Pam Dixon, World Privacy Forum,* at 259.

53. *See In Short Workshop, Remarks of Kevin Trilli, TRUSTe* at 262; *In Short Workshop, Remarks of Jim Brock, PrivacyChoice,* at 274.

54. *See In Short Workshop, Remarks of Sara Kloek, Association for Competitive Technology,* at 243.

55. *See In Short Workshop, Remarks of Prof. Lorrie Faith Cranor, Carnegie Mellon University,* at 273 (calling for platforms to provide hooks for privacy metadata that app developers would be required to supply with their apps).

56. *See In Short Workshop, Remarks of Jim Brock, PrivacyChoice,* at 274.

IV. Post-Workshop Developments

Since the workshop, a number of notable developments have taken place.

First, in July 2012, the NTIA convened a multistakeholder group to develop a code of conduct on how mobile applications should provide transparency for data practices.[57] There have been several meetings since. FTC staff is participating in the multistakeholder effort and is hopeful that stakeholders will consider these recommendations as they work to develop a strong code of conduct. To the extent that strong privacy codes are developed, the FTC will view adherence to such codes favorably in connection with its law enforcement work.

Second, a number of private organizations have provided guidance to app developers on transparency issues through written best practices,[58] boot camps, developer conferences, privacy summits, and workshops.[59]

Third, the California Attorney General ("AG") recently released recommendations for app developers, platform providers, ad networks, mobile carriers, and operating system developers.[60] The recommendations encourage transparency about data practices, limits on the collection and retention of data, meaningful choices for consumers, improved data security, and accountability for industry actors.

Finally, the Government Accountability Office ("GAO") recently published a report in which it recommended that the FTC consider issuing industry guidance on mobile location

---

57. The NTIA multistakeholder process is an outgrowth of the Obama Administration's February 2012 privacy blueprint. This blueprint, among other things, called on: (1) Congress to enact legislation implementing a Consumer Privacy Bill of Rights; and (2) NTIA to convene stakeholders to develop legally enforceable codes of conduct specifying how the Consumer Privacy Bill of Rights applies in specific business contexts. *See* White House, *Consumer Data Privacy in a Networked World: A Framework for Protecting Privacy and Promoting Innovation in the Global Digital Economy* (Feb. 2012), *available at* http://www.whitehouse.gov/sites/default/files/privacy-final.pdf.

58. Future of Privacy Forum and Center for Democracy & Technology, *Best Practices for Mobile Application Developers* (July 12, 2012), *available at* https://www.cdt.org/files/pdfs/Best-Practices-Mobile-App-Developers.pdf; Lookout Mobile Security, *Mobile App Advertising Guidelines* (June 2012), *available at* https://www.lookout.com/_downloads/lookout-mobile-app-advertising-guidelines.pdf.

59. *See, e.g.*, Application Developers Alliance, Privacy Summit Series, *available at* http://devprivacysummit.com; Association for Competitive Technology, *Mobile Industry Unites Behind ACT 4 Apps Education Initiative; Apple, AT&T, Blackberry, Facebook, Microsoft, PayPal, and Verizon Founding Sponsors* (Jan. 9, 2013), *available at* http://actonline.org/act-blog/archives/2782.

60. California Attorney General, *Privacy on the Go: Recommendations for the Mobile Ecosystem* (Jan. 2013), *available at* http://oag.ca.gov/sites/all/files/pdfs/privacy/privacy_on_the_go.pdf.

data privacy.[61] As noted above, the FTC has worked actively to provide such guidance, including by developing and disseminating consumer and business education materials, hosting public workshops and outreach events, and releasing reports on the subject. This report provides additional guidance for mobile companies to consider with respect to disclosing their information collection and use practices.

V. Recommendations

Several workshop participants urged the Commission to set forth clear guidance for businesses on mobile transparency.[62] At the same time, they called on the Commission to avoid being overly prescriptive.[63] Consequently, they advised the FTC to build flexibility into any recommendations or guidelines in order to accommodate dynamic, rapidly evolving technology and new business models.[64]

Commission staff agrees with the participants and accordingly sets forth the following best practice recommendations. These recommendations are organized by industry participant – platforms, app developers, third parties such as ad networks and analytics companies, and app trade associations.[65] The recommendations are intended to be sufficiently flexible to accommodate further innovation and change. To the extent the guidance goes beyond existing

---

61. *See* GAO, *Mobile Device Location Data: Additional Federal Actions Could Held Protect Consumer Privacy*, *supra* note 10.

62. *See Comment of Association for Competitive Technology*, at 9, *available at* http://www.ftc.gov/os/ comments/inshortworkshop/00011-83173.pdf ("Just the same, developers want to continue to innovate but need clear guidance in order to do so."); *Comment of Retail Industry Leaders Association*, at 1, *available at* http://www.ftc.gov/os/comments/inshortworkshop/00003-83138.pdf ("Businesses need clear rules regarding privacy disclosures.").

63. *See Comment of Association for Competitive Technology*, at 10, *available at* http://www.ftc.gov/os/ comments/inshortworkshop/00011-83173.pdf ("The majority of app developers are small businesses that will be unduly burdened if regulations are prescriptive and complicated."); *Comment of Consumer Federation of America and Consumers Union*, at 2, *available at* http://www.ftc.gov/os/comments/ inshortworkshop/00006-83184.pdf ("While it is obviously not the FTC's role to prescribe specific designs . . . .").

64. *See Comment of CMP.LY*, at 1, *available at* http://www.ftc.gov/os/comments/ inshortworkshop/00012-83167.pdf; *Comment of Retail Industry Leaders Association*, at 2, *available at* http://www.ftc.gov/os/comments/inshortworkshop/00003-83138.pdf.

65. Staff recognizes that the mobile ecosystem is complex and involves different relationships among different players, including platforms, such as Apple and Google, hardware manufacturers, such as HTC and Samsung, wireless carriers, such as AT&T and Verizon, and chip manufacturers, such as Qualcomm. These relationships and business models are likely to evolve over time. Although the recommendations in this report are directed towards the predominant business models that exist today, we are mindful that the roles and responsibilities of key players are likely to change over time and, accordingly, the best practices may need to evolve as well.

legal requirements, it is not intended to serve as a template for law enforcement actions or regulations under laws currently enforced by the FTC.

A. Platforms

As noted at the workshop and in the Commission staff's kids' app report, platforms such as Apple, Google, Amazon, Microsoft, and Blackberry are gatekeepers to the app marketplace and possess the greatest ability to effectuate change with respect to improving mobile privacy disclosures.[66]

Specifically, platforms have the ability to set requirements for app developers and even reject apps that fail to meet such requirements.[67] Platforms also control the interface in the app stores and reap significant benefits by serving as the intermediary between apps and consumers.[68] Platforms use the plethora of apps offered on their devices as a significant marketing tool and rely on functionality provided by apps to increase sales of their devices.[69] With the unique position they occupy, platforms could be placing a greater emphasis

---

66. *See In Short Workshop, Remarks of Kevin Trilli, TRUSTe* at 262, 271-72; FTC Staff, *Mobile Apps for Kids: Disclosures Still Not Making the Grade*, *supra* note 28, at 16, fn. 30 ("The app stores set the technical rules that govern what information apps can and cannot access, and provide app developers with a platform to reach consumers. The app developers collect data from their users, and also may integrate functionality from advertising networks, analytics companies, or other third parties that collect data from users. Each participant thus has unique knowledge about the information they are collecting, or enabling others to collect, from users, and must work together to develop disclosures reflecting these practices.").

67. *See, e.g.,* Apple, *App Review Guidelines, Apple Developer* (last visited Jan. 28, 2013), *available at* https://developer.apple.com/appstore/guidelines.html; Google, *Developer Policies, Google Play* (last visited Jan. 28, 2013), *available at* https://play.google.com/about/developer-content-policy.html.

68. David Streitfeld, *As Boom Lures App Creators, Tough Part Is Making a Living*, N.Y. TIMES, Nov. 17, 2012, *available at* http://www.nytimes.com/2012/11/18/business/as-boom-lures-app-creators-tough-part-is-making-a-living.html?pagewanted=all&_r=0 (noting that Apple keeps thirty percent of each app sale); Google Play for Developers Help, *Transaction Fees* (last visited Jan. 28, 2013), *available at* http://support.google.com/googleplay/android-developer/bin/answer.py?hl=en&answer=112622&ctx=cb&src=cb&cb id=-qbqvhfhlgwut (last visited Jan. 28, 2013) ("For applications that you choose to sell in Google Play, the transaction fee is equivalent to 30% of the application price.").

69. Apple, *iPhone, From the App Store* (last visited Jan. 28, 2013), *available at* http://www.apple.com/iphone/from-the-app-store/ ("Hundreds of thousands of endless possibilities…The highest quantity. Of the highest quality. The App Store has the world's largest collection of mobile apps."); Google, *Android Apps on Google Play* (last visited Jan. 28, 2013), *available at* https://play.google.com/store/apps ("Choose from over 500,000 games and apps and your selections will be instantly available on your Android phone or tablet.").

on consumer privacy in their relationship with app developers. Given this background, Commission staff makes several recommendations for consideration by platforms.[70]

1. Platform Disclosures About Information Apps are Accessing from the Application Programming Interface (API)

As noted at the workshop, and in prior Commission staff research, consistency of disclosures is extremely important so that consumers can compare privacy practices across companies.[71] Because platforms have developed a uniform application programming interface (API) through which apps can access standard categories of content on a mobile device, platforms are in a unique position to provide consistent disclosures across apps and are encouraged to do so. Consistent with workshop comments, they could also consider making these disclosures at multiple points in time, as described further below.

i. Just-in-Time Disclosures

First, consistent with the Commission's Privacy Report, before allowing apps to access sensitive content through APIs, such as geolocation information,[72] platforms should provide a just-in-time disclosure of that fact and obtain affirmative express consent from consumers.[73] Providing such a disclosure at the point in time when it matters to consumers, just prior to the collection of such information by apps, will allow users to make informed choices about whether to allow the collection of such information. In addition, platforms should consider providing just-in-time disclosures and obtaining affirmative express consent for collection of

---

70. In its recent revisions to COPPA, the Commission declined to impose COPPA obligations, including the notice and consent requirements, on mobile platforms such as those provided by Apple and Google in their role as providers of access to someone else's child-directed content (they retain COPPA obligations for their own child-directed offerings). Children's Online Privacy Protection Rule, 78 Fed. Reg. 3972 (Jan. 17, 2013) (to be codified at 16 C.F.R. pt. 312). This decision reasonably limited the reach of a prescriptive rule, violations of which may subject a company to significant civil penalties. Here, however, staff is not imposing rules on any members of the mobile ecosystem. Rather, it is identifying areas where ecosystem participants, including mobile platforms, should consider improving mobile privacy disclosure practices.

71. *See In Short Workshop, Remarks of Jennifer King, University of California, Berkeley School of Information,* at 19.

72. *See* Nielsen Wire, *Privacy Please! U.S. Smartphone App Users Concerned with Privacy When It Comes to Location* (Apr. 21, 2011), *available at* http://blog.nielsen.com/nielsenwire/online_mobile/privacy-please-u-s-smartphone-app-users-concerned-with-privacy-when-it-comes-to-location (over half of those surveyed were concerned about privacy when using location-based services or check-in apps); Ponemon Institute, *Smartphone Security: Survey of U.S. Consumers* (Mar. 2011) at 7, *available at* http://aa-download.avg.com/filedir/other/Smartphone.pdf (reporting that 64% of consumers worry about their location being tracked when using their smartphones).

73. FTC, *Protecting Consumer Privacy in an Era of Rapid Change, Recommendations for Businesses and Policymakers*, *supra* note 2, at 60.

other content that many consumers would find sensitive in many contexts, such as photos, contacts, calendar entries, or the recording of audio or video content. Indeed, Apple has used this approach with respect to some of these categories in its iOS6 operating system.

As workshop participants noted, it is particularly important that platforms make these just-in-time disclosures clear and understandable. For example, if an app can access geolocation information over time, the platform should avoid conveying the impression that access is one-time only.[74] Likewise, consumers will better comprehend just-in-time disclosures that avoid technical jargon and use plain language that an ordinary person would understand.[75] As discussed below, consumer testing of various disclosures' effectiveness can help ensure such clarity.

ii. Privacy Dashboard

At the same time, workshop participants noted that a single, just-in-time disclosure may not be sufficient and that multiple disclosures at different points in time may help consumers.[76] For example, as one participant noted, "[H]aving…a lot of different ways and timings for consumers to access those messages, both online and off, is very important."[77] Other design concepts discussed included the need to "create a visual hierarchy" to guide the reader, using concise text, with "clear visual cues to help people find what they need."[78]

These comments suggest that a "dashboard" approach – similar to one used by several existing platforms – may be promising. A dashboard provides an easy way for consumers to determine which apps have access to which data and to revisit the choices they initially made about the apps. There are two different dashboard approaches worth considering – one that is framed by content elements and one framed by applications. Apple has implemented the former approach. With iOS6, Apple has a privacy settings tab which contains entries

---

74. *See In Short Workshop, Remarks of Prof. Lorrie Faith Cranor, Carnegie Mellon University,* at 270-71.

75. *See In Short Workshop, Remarks of Kevin Trilli, TRUSTe* at 252; see also Kleimann Communication Group, Inc., *Evolution of a Prototype Financial Privacy Notice: A Report on the Form Development Project*, *supra* note 37.

76. *Cf.* Cohen, H. H., Cohen, J., Mendat, C. C., & Wogalter, M. S., *Warning Channel: Modality and Media, in* HANDBOOK OF WARNINGS, 123, 123-25 (Michael S. Wogalter ed. 2006), *available at* http://www.safetyhumanfactors.org/wp-content/uploads/2011/12/273CohenCohenMendatWogalter2006.pdf (discussing how conveying a message through more than one modality can be more effective than using a single modality).

77. *See In Short Workshop, Remarks of Pam Dixon, World Privacy Forum,* at 259.

78. *See In Short Workshop, Remarks of Jennifer King, University of California, Berkeley School of Information,* at 16-18.

corresponding to important categories of data such as geolocation, contacts, calendar, and photos. For these categories, users are able to see which apps have access to that piece of information and to turn that functionality on or off on an app-by-app basis.

The second dashboard approach is one that Android has adopted.[79] In the "Settings" menu on an Android device, a user can select "Apps," which provides a list of all apps on the device. By selecting a particular app, the user is taken to a landing page where the user can view information about that app. If a user scrolls down on this page, the user can review the app's permissions, which describe the content that the particular app is able to access on the device. This approach allows for more particularized information on the content that each app accesses.

iii. Icons

Finally, consistent with workshop participant comments, platforms could explore the use of icons.[80] Icons, if appropriately designed and implemented, offer the ability to communicate key terms and concepts in a clear and easily digestible manner.[81] Both Apple and Google utilize icons to signal to consumers when an app is accessing their geolocation information. In Apple's iOS, the geolocation icon appears in the top status bar of a user's device as follows:

---

79. Facebook also offers users a similar dashboard through an applications settings tab where users can see which apps they are using. Users can then select the different apps to get more detailed information about the categories of user information that the apps are accessing.

80. *See In Short Workshop, Remarks of Jennifer King, University of California, Berkeley School of Information,* at 18.

81. *See Comment of Facebook,* at 3-4 *available at* http://ftc.gov/os/comments/inshortworkshop/00009-83165.pdf ("Icons and abbreviations offer practical solutions for achieving simplicity and brevity while also putting consumers on notice of material information."). Facebook further notes that icons have been used successfully in a variety of manners, including on traffic signals, for recycling initiatives, and on remote controls.

On a Google Android device, the geolocation icon is also displayed in the top status bar of a user's device as follows whenever an app is using GPS to determine location:

Consumer testing is important to measure the effectiveness of the disclosures discussed above – whether a just-in-time disclosure, a dashboard, or an icon. As one participant noted, "a key component of the design process is user testing, putting our designs in front of our users to ensure effectiveness."[82] As another noted, "we start creating designs, but we're never right the first time. So, it's a process of iterative test and design where we create, test, refine, create, test, refine."[83]

Companies could test a host of disclosure attributes, such as text, font, and graphics. They could also test what information should appear in disclosures. For example, what categories of information should be included in a dashboard? Where is the balance between providing consumers with relevant information and including so many elements that the dashboard becomes unwieldy or too complex to be useful to consumers? How can industry and others address challenges that exist regarding consumer awareness and understanding of icons?

2. Platform Oversight of Apps

Some privacy practices may not be within the platforms' control. For example, although a platform would know what information the app is collecting through APIs, a platform would not necessarily know what information the app is collecting directly from consumers or what information the app is sharing with third parties. As discussed below, app developers should do a better job of disclosing this information to consumers.

---

82. *See In Short Workshop, Remarks of Jennifer King, University of California, Berkeley School of Information,* at 13.

83. *See In Short Workshop, Remarks of Ilana Westerman, Create with Context,* at 217.

Nevertheless, because of their significant control and leverage over app developers, platforms can play an important role in improving app developer privacy disclosures. For example, consistent with their agreement with the California Attorney General, platforms are already making space in the app store for app developers to provide consumers with information about the app developers' privacy practices. In addition, as our initial kids' app report noted, the platforms should reasonably enforce their contractual requirement that apps have privacy policies.[84]

Further, the platforms could (1) add provisions to their contracts with app developers requiring them to provide just-in-time disclosures and obtain affirmative express consent before collecting or sharing sensitive information; and (2) reasonably enforce these provisions.[85]

Platforms should consider imposing privacy requirements on apps for several reasons. First, many consumers believe that the app stores provide significant oversight of apps; if an app uses their personal information in unexpected ways, this is likely to affect the platform's reputation.[86] Second, including privacy requirements can serve an important education function for small app developers who may not be focused on privacy issues.[87] Third, in many instances, platforms themselves provide app developers with user information; as with all companies that provide personal information to other parties, contractual provisions should address what these third parties can and cannot do with the information.

Finally, platforms could educate app developers on privacy and make available to them important information about consumer privacy considerations as they craft their apps.

These are ideas for further consideration. Commission staff invites suggestions and proposals by platforms for implementing these ideas and looks forward to further discussions

---

84. *See* FTC Staff, *Mobile Apps for Kids: Current Privacy Disclosures are Disappointing*, supra note 28, at 3.

85. The idea that the platforms should reasonably enforce any contractual provisions is consistent with the Commission's approach to privacy and data security generally: a paper exercise alone – such as having written policies and procedures in the back of a file drawer or empty contractual provisions – does not sufficiently uphold the privacy and security of users' information. *See* FTC, *Protecting Personal Information: A Guide for Business* (Nov. 2011), *available at* http://business.ftc.gov/documents/bus69-protecting-personal-information-guide-business ("Your data security plan may look great on paper, but it's only as strong as the employees who implement it. Take time to explain the rules to your staff, and train them to spot security vulnerabilities. Periodic training emphasizes the importance you place on meaningful data security practices.")

86. *See* Jennifer King, *"How Come I'm Allowing Strangers To Go Through My Phone?" – Smartphones and Privacy Expectations* (Draft Under Review) (Sept. 18, 2012) at 6, *available at* http://www.jenking.net/mobile/jenking_smartphone_DRAFT.pdf.

87. *See In Short Workshop, Remarks of Jim Brock, PrivacyChoice,* at 245.

with platforms about how they can better leverage their control over app developers to improve privacy disclosures on mobile devices.[88]

3. Transparency About the App Review Process

Platforms utilize different review processes before offering an app to users through the app store. At least one recent study of a sample of Apple and Android smartphone users found a high degree of consumer confusion among the survey subjects about the nature and extent of review of apps by the platforms.[89] To alleviate any potential consumer confusion, platforms should consider providing consumers with clear disclosures about the extent of review platforms undertake prior to making apps available for download in the app stores, as well as any compliance checks or reviews they undertake after the apps have been placed in the app stores.

4. DNT for Mobile

The December 2012 kids app report noted the significance advertising networks play in collecting data on mobile devices.[90] Because advertising networks often work with multiple developers to provide advertising within apps, advertising networks are in a position to build consumer profiles by collecting consumer data across different applications. Some consumers may not want companies to track their behavior across apps. Indeed, one survey found that 85% of consumers want to have choices about targeted mobile ads.[91]

---

88. These discussions will include, among other things, consideration of the costs and benefits of implementing the recommendations contained in this report. We encourage all key stakeholders in the future to ensure that the agency has ample information on the issues it is considering by participating in the agency's workshops and submitting data-rich comments to inform the discussions. Further, as explained *supra* at note 65, we are mindful of the complexity of business relationships in the delivery of mobile services. Amazon's use of Google's Android operating system in connection with its Kindle Fire device is a good example of this complexity. Although Amazon used Google's open source Android operating system to develop the operating system for its Kindle Fire device, Amazon modified the operating system in significant ways. Unlike other mobile devices that are marketed as Android devices and offer apps through Google's Google Play store, Amazon's Kindle Fire is not branded as an Android device and does not offer the Google Play app store. The relationship between Amazon and Google, therefore, is very different than the relationship between Google and other hardware manufacturers who offer Android devices. Staff would consider Amazon to be the "platform" for the Kindle Fire device but would consider Google to be the platform for Android branded devices. These distinctions and how they may evolve over time offer another reason to invite ongoing discussions about our best practice recommendations for platform providers.

89. *See* Jennifer King, *"How Come I'm Allowing Strangers To Go Through My Phone?" – Smartphones and Privacy Expectations* (Draft Under Review), *supra* note 86.

90. *See* FTC Staff, *Mobile Apps for Kids: Disclosures Still Not Making the Grade*, *supra* note 33, at 6-9.

91. TRUSTe, *Mobile Privacy Survey Results*, *supra* note 12.

A DNT mechanism for mobile devices could address this concern. Accordingly, Commission staff continues to call on stakeholders to develop a DNT mechanism that would prevent an entity from developing profiles about mobile users.[92] A DNT setting placed at the platform level could give consumers who are concerned about this practice a way to control the transmission of information to third parties as consumers are using apps on their mobile devices. The platforms are in a position to better control the distribution of user data for users who have elected not to be tracked by third parties.

Offering this setting or control through the platform will allow consumers to make a one-time selection rather than having to make decisions on an app-by-app basis. Apps that wish to offer services to consumers that are supported by behavioral advertising would remain free to engage potential customers in a dialogue to explain the value of behavioral tracking and obtain consent to engage in such tracking.

Apple has already begun to innovate with a DNT setting on its platform. Apple's iOS6 allows consumers to exercise some control over advertisers' tracking activities via the "Limit Ad Tracking" setting. Although the setting could be more prominent, this is a promising development, and we encourage Apple and other platforms to continue moving towards an effective DNT setting on mobile devices that meets the criteria we have previously articulated for an effective DNT system: that it be (1) universal, (2) easy to find and use, (3) persistent, (4) effective and enforceable, and (5) limit collection of data, not just its use to serve advertisements. We will continue to have discussions with stakeholders in the mobile marketplace on this important issue.

---

92. Industry has devoted a substantial amount of work to developing a DNT mechanism for the web browsing environment. (That ongoing effort would address both desktop and mobile *web* browsing, in contrast to the recommendation here, which would allow consumers to prevent tracking across apps.) Leading browser vendors have developed DNT controls that allow consumers to express a choice not to have behavioral data collected from third-party ad networks or other data collectors. The advertising industry has also taken an active role in this area, developing a standard icon for use by publishers and advertisers to signify behavioral data collection and use and that links consumers to more information and the option to exercise choices about such practices. Most recently, the World Wide Web Consortium (W3C) standard-setting body has worked to develop a consensus standard for a DNT mechanism. *See The Need for Privacy Protections: Perspectives from the Administration and the Federal Trade Commission: Hearing Before the S. Comm. on Commerce, Science, and Transportation,* 112th Cong. (2012) (statement of Jon Leibowitz, Chairman, FTC), *available at* http://www.ftc.gov/os/testimony/120509privacyprotections.pdf.

B. App Developers

App developers also have a critical role to play in informing consumers about their privacy practices.

First, apps should have a privacy policy and make that policy available through the platform's app store. As one participant at the May 2012 workshop noted, this can help a consumer who wants to know what an app developer's privacy policies are before he or she downloads the app.[93] Another participant noted that having a permanent place for a privacy policy can be helpful to consumers after they have begun using a product and have identified a problem or concern they want to research.[94]

FTC staff expects that the California AG's agreement with the leading platforms will accelerate efforts by app developers to create privacy policies.[95] This agreement stipulates that platforms include in the app submission process an optional data field through which an

---

93. *See In Short Workshop, Remarks of Jim Brock, PrivacyChoice,* at 259. This panelist also noted that an added benefit of creating a privacy policy is that it can help educate developers about their own data practices. *See id.* at 245.

94. *See In Short Workshop, Remarks of Pam Dixon, World Privacy Forum,* at 236.

95. Press Release, Office of the Attorney General of California, *Attorney General Kamala D. Harris Secures Global Agreement to Strengthen Privacy Protections for Users of Mobile Applications* (Feb. 22, 2012), *available at* http://oag.ca.gov/news/press-releases/attorney-general-kamala-d-harris-secures-global-agreement-strengthen-privacy. Further, the guidance that the California AG released earlier this month recommended that app developers develop a clear and accurate privacy policy that is conspicuously available to users, including through a link on the app platform page. *See* California Attorney General, *Privacy on the Go: Recommendations for the Mobile Ecosystem, supra* note 60.

app developer can provide a hyperlink to an app's privacy policy, the text of an app's privacy policy, or a short statement describing the app's privacy practices.[96]

Second, app developers should provide just-in-time disclosures and obtain affirmative express consent when collecting sensitive information outside the platform's API, such as financial, health, or children's data,[97] or sharing sensitive data with third parties. The Privacy Report made clear that these categories of information warrant special protection.[98] In the Privacy Report, the Commission indicated that companies should obtain affirmative express consent before collecting or sharing this information, and this recommendation applies equally to app developers. For instance, if an app collects blood glucose information or shares it with third parties, the app developer should provide the consumer with a just-in-time disclosure of that fact and obtain affirmative express consent prior to the initial collection or sharing.

As a general matter, it is important that these app-level disclosures not repeat the platform-level disclosures. For example, an app should be able to rely on the platform's

---

96. Although the agreement has been in place for several months, substantial progress needs to occur. A June 2012 study of 150 of the most popular apps across three leading platforms – Apple's iTunes app store, Google's Play app store, and Amazon's Kindle Fire app store – reveals how much more work needs to take place. *See* Future of Privacy Forum, *FPF Mobile Apps Study* (June 2012) at 4, *available at* http://www.futureofprivacy.org/wp-content/uploads/Mobile-Apps-Study-June-2012.pdf. For example, the study found that only 28% of paid apps and 48% of free apps available in Apple's iTunes app store included a privacy policy or link to a privacy policy on the app promotion page. The top apps in Google's Play store fared even worse. There, only 12% of paid apps and 20% of free apps examined provided access to a privacy policy through the app store. The study did not contain any data on Amazon's Kindle Fire app store because as of the publication of the report, Amazon had not yet provided app developers with the means to comply with the agreement. The Commission staff's kids app reports reached similar conclusions, noting the paucity of information provided to parents before they or their children downloaded popular children's apps. *See* FTC Staff, *Mobile Apps for Kids: Current Privacy Disclosures are Disappointing*, *supra* note 28, at 1; FTC Staff, *Mobile Apps for Kids: Disclosures Still Not Making the Grade*, *supra* note 33, at 4-6. To address this problem, the California AG recently sent warning letters to 100 app developers notifying them that they are not in compliance with California law, which requires the posting of a privacy policy. The developers were given thirty days to conspicuously post a privacy policy within their app that informs users of what personally identifiable information about them is being collected and what will be done with that private information. *See* Press Release, Office of the Attorney General of California, *Attorney General Kamala D. Harris Notifies Mobile App Developers of Non-Compliance with California Privacy Law* (Oct. 30, 2012), *available at* http://oag.ca.gov/news/press-releases/attorney-general-kamala-d-harris-notifies-mobile-app-developers-non-compliance. In addition, the California AG has sued Delta Airlines, one of the recipients of the warning letter. *See* Press Release, Office of the Attorney General of California, *Attorney General Kamala D. Harris Files Suit Against Delta Airlines for Failure to Comply with California Privacy Law* (Dec. 6, 2012), *available at* http://oag.ca.gov/news/press-releases/attorney-general-kamala-d-harris-files-suit-against-delta-airlines-failure.

97. COPPA also requires app developers to obtain parents' consent before collecting personal information from children under 13.

98. *See* FTC, *Protecting Consumer Privacy in an Era of Rapid Change, Recommendations for Businesses and Policymakers*, *supra* note 2, at 59-60.

disclosure that geolocation data will be collected by the app through APIs and need not repeat the same disclosure and consent process. If the app developer decides to share that geolocation data with a third party, the app developer should provide a just-in-time disclosure and obtain affirmative consent from users for that data sharing.[99]

Third, app developers should improve coordination with ad networks and other third parties that provide services for apps so that the apps can provide truthful disclosures to consumers.[100] It is common for app developers to integrate third-party code to facilitate advertising or analytics within an app with little understanding of what information the third party is collecting and how it is being used. App developers should take responsibility for understanding the function of the code they are utilizing.[101] Ad networks and analytics providers are positioned to help app developers with this task. This can only work with greater collaboration and clear and complete communication between the parties.

Finally, as discussed below, app developers should consider participating in self-regulatory programs, trade associations, and industry organizations, which can provide industry-wide guidance on how to make uniform, short-form privacy disclosures.

C. Recommendations for Advertising Networks and Other Third Parties

As discussed, advertising networks and other third parties that provide services for apps should improve coordination and communication with app developers so that the app developers can in turn make truthful and complete disclosures to consumers. Nowhere is this more evident than with respect to the code that ad networks and other third parties supply to app developers to facilitate advertising or analytics within an app. Frequently app developers do not fully appreciate the function of this code, resulting in, for example, an advertising network collecting information without the app developer's knowledge. Ad networks and analytics providers should help app developers better understand how this code works and

---

99. FTC staff recognizes that having a platform disclosure and an app-level disclosure could require two taps from a consumer, and therefore encourage platforms and app developers to develop ways in which the disclosure could be made in one place. For example, the platform could create alternative standardized disclosures that the app developer could pick (*e.g.*, This app collects your location information vs. This app collects and shares your location information.).

100. *Cf.* Children's Online Privacy Protection Rule, *supra* note 70 (noting that app developers will be responsible for third parties that collect personal information from children through their apps).

101. Lookout Mobile Security, *Mobile App Advertising Guidelines* (June 2012), *supra* note 58, at 7-8.

what it does. This can only happen if advertising networks and other third parties improve communications and collaboration with app developers, and vice-versa.

In addition, advertising networks should work with platforms to ensure implementation of an effective DNT system for mobile. This collaboration is important to ensure that consumer choice is honored.

D. Recommendations for App Trade Associations

The trade associations representing app developers also can play a vital role in improving mobile privacy disclosures by developing and improving standardized privacy disclosures, terminology, formats, and model privacy notices. As noted by workshop participants, greater standardization would reduce the potential for consumer confusion and also permit consumers to compare the data practices of different companies and exercise choices based on privacy concerns – similar to how they do so with nutrition labels. App trade associations could improve transparency in several ways.

First, app trade associations could develop standardized icons to depict app privacy practices.[102] One workshop participant unveiled a project still in development – known as the "trust icon."[103] This icon, which appears in the top status bar of a smartphone, signals to a consumer that an app is collecting data by visually bursting three times and then glowing.[104] Users interested in learning more about the data collection can pull down a menu located adjacent to the icon that provides more information about the data being accessed. In addition, a consumer can tap the icon to find out more about the app's data collection and use practices.

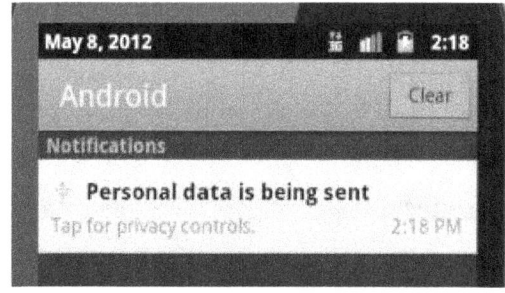

---

102. This report calls on both platforms and app developer trade associations to continue their work to develop icons. This will likely require some cooperation to avoid duplication of efforts or a needless proliferation of multiple icons depicting the same practices.

103. *See In Short Workshop, Remarks of Ilana Westerman, Create with Context,* at 222-31.

104. *See Comment of Electronic Privacy Information Center,* at 6-7, fn. 33, *available at* http://www.ftc.gov/os/ comments/inshortworkshop/00007-83163.pdf (expressing support for forms of notice that utilize visual or aural cues to signal to a consumer that data is being collected).

For each category of data listed (*e.g.*, address book, photos, location), developers could indicate whether they collect or share that information and provide a control to the user, such as an on/off button. In addition, developers could provide information on the "value proposition," or why the app is accessing such information, which according to one participant was crucial to consumers in making informed choices.[105] The development of this icon and disclosure is promising, particularly because it has incorporated feedback from consumer testing.

Second, app developer trade associations could continue work on developing "badges" or other similar short, standardized disclosures that could appear within apps or within advertisements for apps. For example, one participant at the workshop represented Moms With Apps ("MWA"), an online community of developers seeking to promote quality apps for kids and families. She noted that, in response to the FTC's initial kids app report, MWA created a privacy badge icon available at no cost to app developers.[106] The badge is designed to be featured in the preview screenshots in the app store that a consumer can review prior to downloading an app, in an app itself, on an app developer's website, or on a social media page.[107]

The MWA badge, which combines icons and limited amounts of text (*e.g.*, "No Ads" or "Has In-App Purchases"), was originally conceived to provide information to parents about five areas: (1) whether an app collects or shares data; (2) whether an app contains advertising; (3) whether any purchases can be made within the app; (4) whether an app shares information with social networks; and (5) whether an app includes external links to other websites.[108] Since the workshop, the badge has undergone additional refinement and now includes additional data fields, such as the recommended minimum age for an app.[109] Commission staff encourages this kind of innovation. Staff also recommends that any exploration of privacy badges include consumer testing to assess their ability to communicate privacy practices clearly and accurately to consumers.

---

105. *See In Short Workshop, Remarks of Ilana Westerman, Create with Context,* at 225.

106. *See In Short Workshop, Remarks of Sara Kloek, Association for Competitive Technology,* at 241 (representing MWA).

107. *See* PRWeb, *Parent App Developers Announce New Privacy Disclosures for Users* (Feb. 29, 2012), *available at* http://www.prweb.com/releases/2012/2/prweb9229537.htm.

108. *Id.*

109. *See* eWeek, *ACT Launches App Privacy Icons* (Oct. 4, 2012), *available at* http://www.eweek.com/developer/act-launches-app-privacy-icons; App Trust Project, *available at* http://apptrustproject.com (displaying sample badge).

Third, app developer trade associations could develop ways to have more standardization within app privacy policies.[110] At the workshop, PrivacyChoice and TRUSTe discussed their respective tools to generate privacy policies optimized for a mobile device that are easy for consumers to read and understand and at the same time easy for mobile app developers to create. With both tools, developers answer a series of standard questions about their app, such as what types of data it collects, whether and with whom it shares data, and whether a consumer can delete the data collected. The questions drive at the heart of the most important practices of interest to consumers (*e.g.*, use of location information, targeted advertising). For app developers, using either generator may sensitize them to privacy issues and require them to take stock of what their apps are doing with the data they collect.

Based on the responses provided, the generators create a standardized, layered policy that eliminates long, complicated policy language in favor of icons and streamlined disclosures presented in a short sentence (*e.g.*, "We don't share personal information with marketers.") or phrases (*e.g.*, "Location Services" or "Tracking Technologies"). This initial interface allows consumers to view the highlights of the policy at a quick, initial glance. For consumers desiring more information, they can touch any of the icons or phrases and drill down one layer where more detailed, granular information is available. Again, Commission staff supports this type of innovation as a way to provide a starting point for improved disclosures.[111]

As these efforts to create standardized icons, badges, and privacy policies intensify, Commission staff recommends consideration of three additional steps.[112] First, staff encourages trade associations to take into account the important work of academics,

---

110. The Commission has supported standardization of disclosures in other contexts. *See* Kleimann Communication Group, Inc., *Evolution of a Prototype Financial Privacy Notice: A Report on the Form Development Project*, *supra* note 37, at 273-74 ("Standardization of form and content helped consumers recognize the notice and the information in it. As they became familiar with the prototype, they learned where to look for the differences. Standardization reduces cognitive burden because consumers recognize the information without having to continually re-read notices word for word.").

111. The Commission staff's study on mortgage disclosures highlighted the importance of layered disclosures. That study compared existing mortgage disclosures to a prototype disclosure form developed by agency staff. The prototype form included summary information on the first page, and additional information on pages two and three. The prototype form significantly improved consumers' recognition of mortgage costs. *See* FTC Staff, *Improving Consumer Mortgage Disclosures: An Empirical Assessment of Current and Prototype Disclosure Forms*, *supra* note 39.

112. These recommendations – such as creating standardized, model short form privacy notices, soliciting input from academics and other experts, and utilizing consumer testing to assess the effectiveness of various disclosures and ensure meaningful consumer comprehension – build on ideas and concepts advanced in the NTIA process and elsewhere. *See, e.g.,* National Telecommunications and Information Administration, *Privacy Multistakeholder Meeting Final Agenda* (Jan. 17, 2013), *available at* http://www.ntia.doc.gov/files/ntia/publications/agenda_1-17-13_final.pdf.

usability experts, privacy researchers, and others in developing icons and other standardized approaches. Staff also encourages continued work in this area by these groups. Second, as noted above, companies should consider consumer testing of new mechanisms to ensure meaningful consumer comprehension. Third, any standardized icons, terminology, format, privacy notices, or other disclosures should be accompanied by a robust education campaign.[113]

At the same time, Commission staff is mindful of problems that could result from a proliferation of icons, badges, and notices. As noted at the workshop, approaches that rely on icons, short disclosures and other methods are unlikely to succeed without some degree of uniformity, as consumers may have difficulty recognizing multiple icons and forms of short disclosures.[114] Accordingly, Commission staff urges stakeholders to come together to develop complementary and consistent approaches. The NTIA multistakeholder meetings may be one venue in which such complementarity and consistency can be achieved.

Finally, as this work continues, FTC staff continues to encourage trade associations to educate app developers about information collection and use practices. One participant noted that app trade associations are educating app developers about privacy issues through boot camps, workshops, panels, and other activities.[115] Participants discussed that this type of education should focus not just on transparency issues, but also on how to build privacy protections into apps by employing privacy by design.[116] Commission staff agrees.

VI. Conclusion

Mobile technology benefits consumers through innovative content, products, and services. At the same time, mobile devices raise a number of potential privacy risks. Informing consumers of these risks through meaningful and effective disclosures presents significant challenges that the FTC explored in its May 2012 workshop. This report distills the Commission's prior work on these issues, along with panel discussions and written submissions, to develop a number of recommendations for companies to improve mobile

---

113. *See In Short Workshop, Remarks of Prof. Lorrie Faith Cranor, Carnegie Mellon University,* at 237-38.

114. *See In Short Workshop, Remarks of Prof. Lorrie Faith Cranor, Carnegie Mellon University,* at 255.

115. *See In Short Workshop, Remarks of Sara Kloek, Association for Competitive Technology,* at 241-42.

116. *See In Short Workshop, Remarks of Pam Dixon, World Privacy Forum,* at 256-57; *In Short Workshop, Remarks of Jim Brock, PrivacyChoice,* at 274.

privacy disclosures. These recommendations are intended to guide companies as they continue to create exciting and promising products and services for consumers.

Looking ahead, however, many questions remain, including the following: What information should be included in app developer privacy policies? What might a model short privacy notice look like? Can a single system of icons be developed to avoid consumer confusion? NTIA's multi-stakeholder process is focusing on how mobile transparency can be improved, and is well positioned to address some of these questions

In the meantime, FTC staff strongly encourages companies in the mobile ecosystem to work expeditiously to implement the recommendations in this report. Doing so likely will result in enhancing the consumer trust that is so vital to companies operating in the mobile environment. Moving forward, as the mobile landscape evolves, the FTC will continue to closely monitor developments in this space, including evolving business models, and consider additional ways it can help businesses effectively provide privacy information to consumers.